The Grand Experiment

Hassan Rasheed

ISBN
978-1-105-01426-0

Printed in the United States of America
By Lulu.com

Chapter 1: The Hatchlings

First officer of the galaxy ship **Uweif** of the **nnevjfd** federation said, "Eject (translation from the alien language of **Mathora**)"

Second officer pulled a lever and a pod with its two frozen-in-time seeds were ejected from the space ship's belly and started its descent to the planet Ard. Ard is a red-Yellow-ish planet reflecting the iron that is plentiful on its surface's sand and rocks. The descent took about 49 minutes and ended with the pod extending 11 landing arms cushioning the final break in speed as it touched down.

Ard is the third planet orbiting the star QRT103 in the Rusty Riggins galaxy which is 300 million light years away from our Milky Way. Ard is similar to Earth in size and the amount of energy it received in the form of light from QRT103. It is a lifeless planet which was exactly what the **nnevjfd** alien life forms wanted for they had plans of starting new life on its sterile surface. The **nnevjfd** were known as The Creators of life in the Universe.

After the pod settled down on Ard's surface for about 2 hours a metallic beam started to immerge from its top and after reaching 302 feet (translated from the alien **nitr**) a whirlymajig thing unfurled that was about 53 feet in diameter and started to whirl around propelled by the wind providing the energy required to liven the pod and the top of the pod was lifted up on 5 pillars.

The Grand Experiment

The two seeds were then ejected out of the pod landing next to it with tubes connecting them to its power supplies. After 54 minutes (translated from the alien **ϟϼϾϒ**) the red seed snapped open oozing a red slime and a red entity too horrid to describe emerged. After another 13 minutes the green seed snapped open emitting the smell of a dead corps and another horrid entity emerged green in color. The two entities then hugged the only way they could and then sat down taking in the vast desert looking land scape with the distant Sakara mountain silhouetted against the red sky as QRT103 set behind it.

(Please note: From this point forward, I will not include the original alien text and the reader will assume the translation of expressions, postures and emotions have already been made to convey the closest approximation to human expressions, postures and emotions.)

Chapter 2: The Assignment

As QRT103 rose in the sky the next day warming up the cold-blooded aliens they started to move around and stumbled onto a plaque with the following inscription:

"The Federation of Zulficar High Command's

Right Of Passage Instructions for Citizenship:

You are to bring forth life on this barren planet called Ard.
You are to produce a community of one billion species living under the irrefutable laws of nature to a state of pure stability.
Only then will you be allowed to be citizens of the federation and return to the planet Zanab where you will choose a mate and have the opportunity to prosper and pursue happiness."

The Grand Experiment

Chapter 3: Gurdo and Norweedo

The revolting red alien entity was known as Gurdo whom we can call the male while Norweedo was the name of the ghastly green alien entity whom we can assumed to be the female. Sex had nothing to do with these two aliens which some of you may assume by the sexual designations but they had sex organs that did not fit. The assignment by the high command to the two of them was to ensure the citizenship of the federation were capable individuals. The only intercourse between red Gurdo and green Norweedo was that of solutions to problems, ideas, plans and the execution of their ideas and plans.

Gurdo said, "I think we need to start from the top and work our way down in the design process. What do you think, Norweedo?"

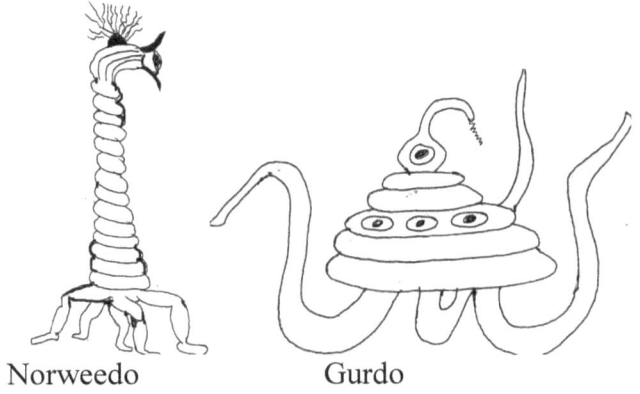

Norweedo   Gurdo

Norweedo thought for a few seconds and then replied, "I don't think so! Things will get more complicated as we move down to more details. I think we need to start from the bottom and work our way up."

"I disagree!", said Gurdo as he pounded one of his scab riddled tentacles on the ground.

"And I disagree with you!" insisted Norweedo as she made noises with her hairy head knob.

Gurdo reached over to Norweedo with one of his tentacles, grabbed her by her eye stalk and started to pound her into the ground so Norweedo grabbed Gurdo's testicles and started to pull so the fight ended in a hurry.

As the dust settled, Norweedo asked Gurdo why he wanted to start at the top of the design process and he responded and said it was because if we know what the end result should be we can work from that point and work out the details later. Norweedo then said she really did not know what the end result Gurdo was talking about. All she knew was that what they create needed to be alive and viable.

"That is exactly what I mean. We take a look at this planet and decide how life should function to survive here and design life to suit it," responded Gurdo.

"But we know very little about this planet," responded Norweedo. "It will take a billion years of study before we can even start the design process."

"What are you complaining about?" exclaimed Gurdo. "We have all the time we need. If it takes a billion years then so be it. No skin off of your nose, Norweedo!"

"What if it takes 2 billion years, another 2 billion years to design while the star QRT103 only has 6 billion years left before it collapses. That is not much time to test out our life form let alone let it live and evolve. We will have failed our mission," responded Norweedo.

"Oh, you are so pessimistic, Norweedo. It will not take that long!" insisted Gurno.

"Can you guarantee that, Gurdo?" asked Norweedo.

Gurdo's top eye started to get blood-shot from the tension and so he reached over to Norweedo with one of his scabby tentacles, grabbed her by her eye stalk again and started to pound her into the ground so Norweedo grabbed Gurdo's testicles again and started to pull until the fight stopped.

"You can't have a tantrum every time we disagree, Gurdo," said Norweedo. "As far as I am concerned, we need to start small and let nature take its course."

"What do you mean?" enquired Gurdo.

"I mean, we should select a microcosm of this planet, design life around it and then let it go to evolve on its own. I bet you, once it gets started on the right path it will take no time at all to be a success and this planet would be transformed," responded Norweedo.

The Grand Experiment

Chapter 4: Entropy

After some convincing, Gurdo agreed to the concept of starting from the bottom working up instead of the other way around. That did not mean their work would be anything like a walk in the park. They still had to decide on a natural law to follow in order to make life possible and its ability to evolve.

After a considerable number of arguments where Gurdo failed to control his temper causing a sever black eye to Norweedo, they decided to try the concepts associated with the natural law of Entropy which states "Entropy is the degree of disorder or randomness in a system". They realized life had a high degree of order and therefore had a low entropy value. They figured a chemical reaction is the simplest way to try out this concept. In order to decrease the entropy value of a chemical compound, they had to provide a way of increasing entropy elsewhere in the system perhaps in another chemical compound.

They imagined a chemical compound consisting of three atoms; Blue, Orange and Yellow in color. We will call this compound the BOY compound. They also imagined many copies of the BOY compound floating in a pond where other atoms and compounds existed. Specifically, there were many copies of two other independent atoms named Gray and Green (G&G) respectively floating in this pond. The BOY compound had the ability to attracted and allow the bonding of the G&G atoms while its atoms were released from forming its compound in the pond.

The Grand Experiment

When those three BOY atoms were released randomly in the pond, they increased the pond's randomness and therefor increased the Entropy value by a value of 3. Meanwhile the two G&G atoms that were bonded together in turn decreased the pond's Entropy by the value of 2 the end result being the net increase of the Entropy of the pond by the value of 1 (3-2=1). Norweedo and Gurdo then imagined bonding process continuing with other G&G atoms creating a chain and then many chains in the system that would form a sort of spherical membrane. This membrane then becomes the first element in a very primitive life form we will call the GG species.

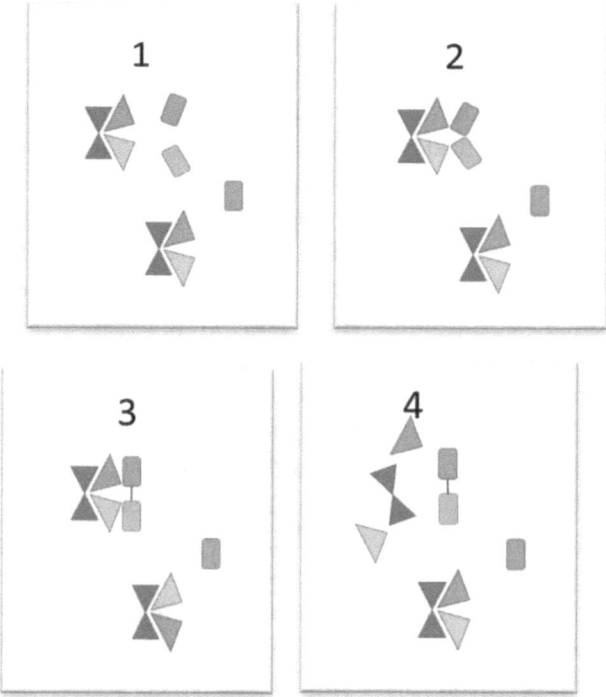

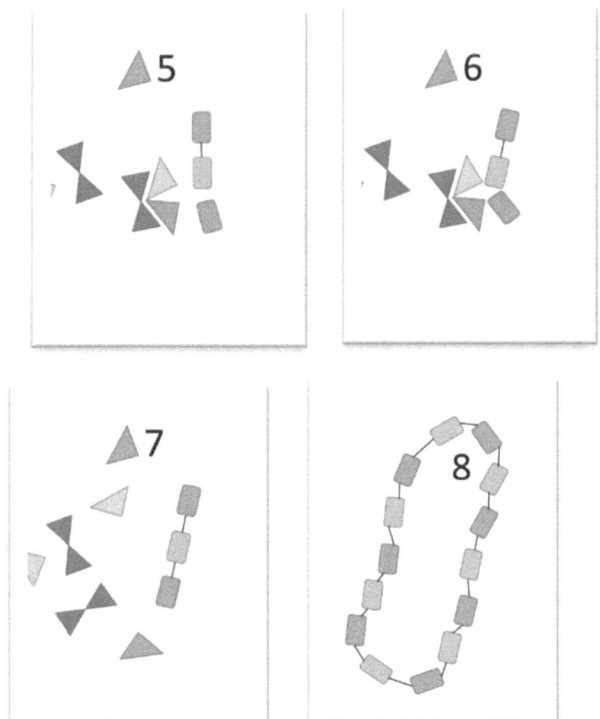

Sequence of life's evolution plan on paper.

The Grand Experiment

Chapter 5: The Experiment in a Bucket

The next step in their quest to create life was for Gurdo and Norweedo to find the most appropriate atoms and compounds to mimic what they had drawn on paper. It took 0.3 billion years to find what they were looking for. They found a chemical compound that acted in a similar way to the BOY compound which brought other atoms together and helped them bond creating a new chemical compound.

The only difference was that the BOY type chemical compound consisted of 6 atoms which Gurdo and Norweedo called the BOY-6 chemical compound and which they found in a pond called the Osta-Bay-Pond. This BOY-6 chemical compound helped 3 other atoms they called the 3 GG atoms bond together to form a chemical compound they dubbed GG-3. they had found the 3 GG atoms in the Glaptos-Island-Pond on the other side of the Ard.

Then they had to find a way to easily and quickly detect when the GG-3 compounds were generated in the solution. Gurdo wanted to create a nuclear microscope to look for the GG-3 compounds while Norweedo insisted on a simple strip of paper with a reactive chemical that would change color when the GG-3 compound was present. This caused Gurdo to have a tizzy fit. Norweedo quickly curled up around her eye causing Gurdo to start to bounce her around like a basketball. In the end Gurdo got his way.

Unfortunately, when they put the solution of BOY-6 compounds in a bucket with the 3 GG atoms Gurdo could not find any evidence of the GG-3 compound forming in the solution. "Here, let me take a look. Your eyes must be going bad," said Norweedo as she took over the atomic microscope from Gurdo. But she could not find any evidence to the GG-3 compound either.

They spent the next hundred million years trying to figure out why nothing was happening after which time they discovered they were putting the bucket containing the mixture in the shade of a hill. When they moved it out of the shade under the rays of QRT103 and in a darker colored bucket they started to detect the GG-3 compounds in the solution. They also detected that the BOY-6 chemical compound did not break apart when the 3 GG atoms came together. It only changed shape.

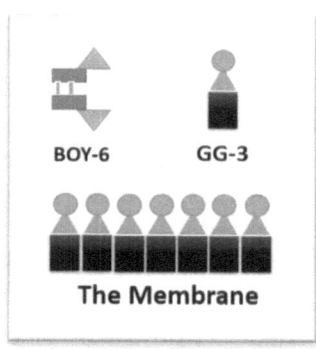

Structure of Sapphire

To their surprise and as the number of GG-3 compounds increased in number they formed spherical membranes containing the reactive BOY-6 and 3 GG atoms which then formed more GG-3 compounds that were then added to the membrane. As the membrane grew it would rupture forming smaller spherical bodies that then grew and repeated the process. At this point and in a rare moment Gurdo and Norweedo agreed they had caused the formation of the first species on the planet Ard and they named it Sapphire.

Hassan Rasheed

The Grand Experiment

Chapter 6: The Evolution of Sapphire

From the BOY-6 and GG-3 compounds more complex compounds were generated and eventually an entity was able to reproduce itself which was the most complex compound to that point in their experiment. It did not take them long to realize that an increase in complexity was the character selected for by nature resulting in the Sapphire species.

Indeed, during the next 3 to 4 billion years the complexity of species increased. That is not to say less-complex species did not survive. They did because they found special environments that suited them just right and gave them an advantage over more complex species.

By the end of the 4 billionth year, Ard was a collage of species large and small, complex and less so. Some species became the environment to others while some became the nutrient base for the survival of others. From the beginning, energy was a key factor in the whole experiment.

Entropy of Ard meanwhile kept increasing exponentially up to the end of the 3 billionth year where it leveled off considerably due to the fact the living were made of relatively soft tissue and they could not penetrate far into the surface of Ard. But that was OK. Life continued for another billion years scratching the surface of Ard and recycling its nutrients.

The Grand Experiment

Chapter 7: Packing Up

As the population of one billion species was being approached on the planet Ard the two aliens started to pack up for the trip back to the planet Zanab and rejoin the federation of Zulficar once again to find a mate and the opportunity to prosper and pursue happiness. It is true Norweedo came up with the idea of starting from the bottom up and letting nature take its course would be the delight of the federation, Gurdo felt he would be left out and get less recognition.

After Norweedo completed his packing, he decided to take a nap. Gurdo on the other hand obsessed over his prospects. He thought he would get less opportunities. That he would not find the ideal mate spending the rest of his life in misery pursuing a life of mediocre breaks driving him to an early death. He thought he had to come up with a plan to improve his chances.

The Grand Experiment

Chapter 8: The Apple and The Log Ladder

Gurdo brooded about what he should do as he traveled the mountains and valleys of Ard. When he stopped to have a drink of water from lake Karoon he heard a loud rumble and the ground started to shake. He looked around and saw a huge cloud of dust approaching and realized it was a herd of Hahunas stampeding towards him. He did the only sensible thing and dove into the lake to take cover.

The Hahunas were a species of ground dwellers that were 98 feet tall on average weighed around 30 tons a piece and looked like a very large elephant with flappy ears but no hind quarters. They balanced themselves on their hands to walk and run holding their bellies up in the air.

It turned out they, the Hahunas, were thirsty and were headed to the lake to take a drink. One of them sneezed after taking in some water and out came Gurdo flying in the air and landed on a Malony tree top. He stayed there until the herd settled down for a nap next to a grove of apple trees. He then got down from his perch and started to think. He looked at the resting Hahunas then looked at the apple trees and came up with an idea.

He dressed up like a snake and slithered down one of the apple trees and stood close to the ear of one of Hahunas and said, "Hahuna want an apple?"

To Hahunas, apples were a delicacy that were also out of their reach being 99 feet high on average so it was not hard for the Hahuna in question to pay attention and get up. "Who is there?" he asked the Hahuna as it looked around.

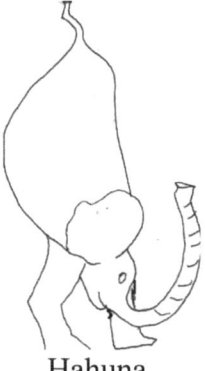

Hahuna

"psst, I am down here," yelled Gurdo, "Hi I am Gurdo. What is your name?"

The Hahuna looked down and saw the snake and said, "I am Sakanuts soon to be father of Sakabolts and leader of the Sakaknobs tribe. How can I help you?"

"I am here to help you!" replied Gurdo.

"Help me? I need no help! I am the leader of the Sakaknobs tribe. I got all the help I need!" the Hahuna replied.

"Do you want to reach some of those apples?" asked Gurdo as he pointed to the apples just out of reach of Sakanuts' snout.

"Sure, I do! Who wouldn't?" responded Sakanuts as he looked up at the apples while his mouth drooled onto Gurdo.

"You will have to become intelligent to get to the apples," replied Gurdo as he wiped the drool from his face.

"How do I become intelligent?" asked Sakanuts.

"You will have to start using tools," replied Gurdo.

"What is a 'tool'?" asked Sakanuts.

"See that log over there by the lake?" replied Gurdo.

"Yes, I see it," replied Sakanuts.

"How high is it from the ground?" asked Gurdo.

"It looks like it is 2 feet high," replied Sakanuts.

"And how far away from your snout are the apples?" asked Gurdo.

"About a foot," replied Sakanuts.

"So go bring that log over here under the apple tree," said Gurdo.

To a 98 foot Hahuna a log with a diameter of two feet was but a twig and so he brought it over put it under the apple tree.

"Now stand on it," commanded Gurdo of Sakanuts and as the latter stood on the log, he was able to reach the low hanging apples.

As word got around the population of Hahunas, log ladders were being created and used to reach not only apples but also other delectable foods growing on trees. Gurdo then introduced them to the idea of transferring the concept of tool making to other situations such as killing mouthwatering flying species as well as ground and water dwellers that were also delicious.

The Grand Experiment

.

Chapter 9: The Confrontation

When Norweedo woke up from her nap she found all these log ladders scattered around. She knew they were not natural results of evolution. She walked over to Gurdo holding a log ladder and asked, "What is this?"

"Isn't it neat! I helped the Hahunas build it. It was amazing they took the original concept and expanded on it," answered Gurdo.

"But it doesn't look natural, Gurdo," explained Norweedo.

"What do you mean it doesn't look natural," complained Gurdo.

"It doesn't look like something evolved naturally on this planet which was one of the conditions of our mission statement!" Responded Norweedo.

"I don't see why not after all a Hahuna created it and Hahunas are natural products of evolution!" mumbled Gurdo.

"Yes, Hahunas are natural but unfortunately their brains are limited in what they can think!" answered Norweedo.

"What do you mean?" asked Gurdo.

"Their brains depend on their physical sensors. These sensors are limited in what they can sense from their environment and in turn their brains will produce limited results, questionable quality and accuracy," responded Norweedo.

"I don't see anything wrong with that. The Hahunas are the result of evolution and anything they do must be considered natural after all they are the products of nature," added Gurdo. "The Hahunas were so happy they called this the Intelligence Effect."

"But they should not be able to create anything besides images from what they see and music from what they hear. This thing you helped them create was not an image of anything nor did it make a sound familiar to any sound on Ard," Responded Norweedo.

"I don't see the harm in creating something functional by the Hahunas. They have every right to pursue wealth and happiness. After all that would be a feather in our cap to show how advanced the Hahunas have become," complained Gurdo.

"No good will come of it, Gurdo. You can't mix the laws of the universe with inaccurate and anemic rules that Hahunas come up with," responded Norweedo. "Your Intelligence Effect will only cause hardship for all!"

Chapter 10: The Warning

Frustrated with Gurdo, Norweedo approached the Hahunas in an effort to dissuade them from the pursuit of the path of eminent destruction. She sees one laying down with his flappy ear close to the ground and approached. "Good mid-day my dear fellow. I am Norweedo your local voice of reason. Whom might you be?"

The Hahuna looked around and asked his partner, "Please pass me the fly swatter."

The partner grabbed it from the picknick basket and handed it to him.

The Hahuna took a swipe at Norweedo but missed. Norweedo jumped quickly under a rock to take cover and stayed there until the Hahuna set the swatter down. Norweedo then crawled out and approached the Hanuna once again, "Good mid-day my dear fellow. I am Norweedo your local voice of reason. What is your name?"

The Hahuna reached again for the swatter but before he could swipe Norweedo yelled, "Don't do it you crazy son of a bitch. I am here to help you survive on this Ard you idiot!"

"Who is that I hear?" asked the Hahuna of his partner.

"I don't hear anything!" responded the partner.

With both eyes now open, the Hahuna sat up and responded with, "Who is there?" as he looked around.

"Good mid-day my dear fellow. I am Norweedo your local voice of reason. What do Hahunas call you?"

asked Norweedo as the Hahunas wagged his large flappy ears.

"Oh, there you are!" said the Hahuna as he spotted Norweedo. "I am Sakanuts soon to be father of Sakabolts and leader of the Sakaknobs tribe. How can I help you?"

"I am here to help you and your tribe." said Norweedo. "You see the Intelligence Effect given to you by Gurdo gave you the wrong impression that intelligence was what you needed to succeed on this planet. It really is a deceptive way of life."

"How so?" asked Sakanuts.

Norweedo responded and said, "The Hahuna brain is limited..."

Sakanuts interrupts Norweedo abruptly and said, "Are you saying we are retarded?"

"No, I am not. Let me explain for a minute and you will understand!" replied Norweedo.

"Go ahead then!" said Sakanuts.

"All brains have basic limitations that are twofold. The first is their physical limitation in storing information and the second is the limitations of its sensory organs. For example, the eye can only see a limited spectrum of light which is called the visible spectrum. The same type of limitations exists for time, sound, touch, smell, and hearing. Therefore, the brain is not able to detect or store the whole spectrum of reality," continued Norweedo. "Although our brain's sketches are greatly reduced in information, it is a powerful tool. The brain is able to manipulate its sketches and produce new ones. For example, the brain records a sound one day and another the next day and perhaps combine them on the third day to produce a completely new melody.

The above demonstrates how to solve problems. By bringing together two different sketches and producing a third may solve a need. Unfortunately, since the sketches are limited in information and the process of bringing together two sketches does not follow any natural laws. The end result is an artificial solution that will not work in the real world."

"Wow, hold on there. What are you talking about?" questioned Sakanuts.

"Say your brain can store 50% of the information that is out there and your sensory organ can process only 50% of what comes its way then the total information is 25% of reality," responded Norweedo.

"I gotcha now. Carry, on!" said Sakanuts.

"If the artificial world that the Hahuna brain creates is in the form of square blocks it is doomed to fail because reality is round. To build a square structure on a round object like a ball is unstable," continued Norweedo.

"You are now making some sense to me!" said Sakanuts. "Unfortunately, Ard is flat and we can build squared log ladders that work just fine to help us get to the apples."

"You can't make a log square. It is round and unstable. It will roll," said Norweedo.

"We got around that problem. We simply set two logs on the ground and then set two more cross wise and tie them all with supple tree branches," responded Sakanuts.

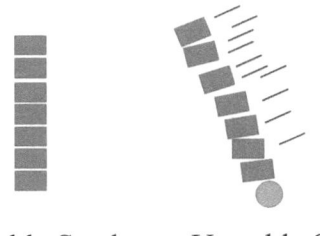

Stable Stack   Unstable Stack

"Just remember to let nature build things for you because then they will be more stable and durable than anything you can build. There is a very good reason why apples are out of reach for you. In whatever you do let nature set the path because nature is the final authority on what is right and suitable," concluded Norweedo.

Chapter 11: To Grow and Prosper

All this time, Gurdo was watching what was happening between Sakanuts and Norweedo from behind a boulder. After Norweedo left the scene, Gurdo put on his snake suit and slithered over to Sakanuts and whispered in his ear, "Norweedo is the devil incarnate and is not to be trusted. I have given you the righteous path to grow, prosper and multiply." Sakanuts took that to heart and spread the word among his kind.

From that point forward and for a measly 150 years the Hahunas sure did multiply and prosper. They invented every tool imaginable including QRT103 ray capturing panels that supplemented their insatiable need for energy.

The Grand Experiment

Chapter 12: The End Result

The Hahuna populations grew and grew and grew to the point these large flat ray capturing panels covered the whole Ard preventing the rays of QRT103 from reaching any other creation including the apple trees. This caused mass extinctions but the Hahunas did not care for they were able to make their own artificial apples. Their abilities of manufacturing anything artificial was quite efficient and they had no need for the natural.

Unfortunately, the byproducts of manufacturing of the artificial caused all micro-organisms to die off. As the trees died due to the lack of light so did the other species that depended in them for sustenance such as the flying species and ground dwellers. With the absence of micro-organisms, bodies of the dead started to pile up because they did not break down. The Hahunas did not care because the bodies did not stink and they simply tossed them in the lakes where they sank out of sight.

Then there was the problem of their excrement which also did not stink but was uncomfortable to step into so they invented what looked like shovels and they got rid of the shit by shoveling it into the lakes as well. But eventually the lakes were full of bodies and filth and they had a hard time finding sources of clean water. They figured there was no harm in drinking lake water that did not stink and sure enough no one got sick while their intestines adjusted to the new or should I say the familiar.

The Grand Experiment

Chapter 13: The Price to Pay

The leader of the federation caught wind of what was happening on the planet Ard and dispatched a team of rectifiers to see if they could salvage anything from the mess created by Gurdo and Norweedo. When they arrived, they found them fighting over who was to blame for the Ard apocalypse. The rectifiers told them they were supposed to work as a team and therefore they were both responsible.

The rectifiers huddled to discuss what to do with these two disgraceful beings. After about an hour the team decided that both were to hunt down and kill all the Hahunas since they were compromised by unwise thoughts of grandeur and were beyond hope of accepting nature as their guide in life. They were then to start populating the Ard with a billion new species all over again.

As QRT103 was setting at the end of the day, the two aliens started to argue again about the approach they will take to accomplish their task. "I advocate the top-down approach," said Gurdo.

"It will take billions of years just to figure out your design. I insist we take the bottom-up method and let nature decide what that design should be," countered Norweedo.

The Grand Experiment

Hassan Rasheed

The End

www.ingramcontent.com/pod-product-compliance
Lightning Source LLC
Chambersburg PA
CBHW061232280526
45784CB00006B/2733